Logistic and Multinomial Regressions by Example
Hands on approach using R

First Edition

I0511234

Faye Anderson, MS, PhD

Contents

Veritas

Preface

During my early college years, I tried to avoid using logistic regression because of all the negative rumors about its difficulty. Thanks to my professors and their applied approach, I let go of my *logit, probit,* and *multinom* fears! The purpose of this book is to help others understand the beauty and ease of conducting such regressions whenever needed.

The best way to benefit from my book is to do the examples and read the interpretation. The following table summarizes the types of models in each example:

Regression/Topic	Number of response values	Number of predictors	Examples
Univariate logistic	2	1	1, 3
Probit	2	1	2
Multivariate logistic	2	More than 1	4
Multinomial univariate	>2	1	5
Multinomial multivariate	>2	More than 1	6
Ordinal	4	1	7

As with my other books, R was selected because of its free accessibility and assumes limited knowledge of the software. If you have any questions please feel free to post them in my Amazon author's blog.

Enjoy!

Chapter 1: Logistic Regression

Logistic regression (also known as binary logistic regression) predicts the dependent variable (Y) using a set of independent variables (predictors) just like ordinary linear regression but the dependent variable in this case is binary. That is, it only takes two values.

The odds ratio (OR) equals the probability of one outcome over the probability of the other (P(Y=0)/P(Y=1) = p/q; q=p-1). The natural log of odds is called the logit, or logit transformation, of p: logit(p) = loge(p/q).

Example 1: Univariate Logistic Regression

This example uses Motor Trend Car Road Tests (mtcars). We fit a model to predict the probability of a vehicle having a V engine or a straight engine given its weight (in 1000 pounds). After interpreting the model outcomes, we assess the goodness of fit for the model and conclude with ROC curve.

```
> data(mtcars)
> head(mtcars)
                   mpg cyl disp  hp drat    wt  qsec vs am gear carb
Mazda RX4         21.0   6  160 110 3.90 2.620 16.46  0  1    4    4
Mazda RX4 Wag     21.0   6  160 110 3.90 2.875 17.02  0  1    4    4
Datsun 710        22.8   4  108  93 3.85 2.320 18.61  1  1    4    1
Hornet 4 Drive    21.4   6  258 110 3.08 3.215 19.44  1  0    3    1
Hornet Sportabout 18.7   8  360 175 3.15 3.440 17.02  0  0    3    2
Valiant           18.1   6  225 105 2.76 3.460 20.22  1  0    3    1
> help(mtcars)
```

Description

The data was extracted from the 1974 Motor Trend US magazine, and comprises fuel consumption and 10 aspects of automobile design and performance for 32 automobiles (1973-74 models).
Usage
mtcars
Format
A data frame with 32 observations on 11 variables.

```
[, 1] mpg    Miles/(US) gallon
[, 2] cyl    Number of cylinders
[, 3] disp   Displacement (cu.in.)
[, 4] hp     Gross horsepower
[, 5] drat   Rear axle ratio
[, 6] wt     Weight (lb/1000)
[, 7] qsec   1/4 mile time
[, 8] vs     V/S
[, 9] am     Transmission (0 = automatic, 1 = manual)
[,10] gear   Number of forward gears
[,11] carb   Number of carburetors
> summary(mtcars$wt) # car weight in 1000 lb)
   Min. 1st Qu.  Median    Mean 3rd Qu.    Max.
  1.513   2.581   3.325   3.217   3.610   5.424
> summary(mtcars$vs) #engine type v or s
   Min. 1st Qu.  Median    Mean 3rd Qu.    Max.
 0.0  0.0000  0.0000  0.4375  1.0000  1.0000
> table(mtcars$vs)

 0  1
18 14

plot(mtcars$wt, mtcars$vs, xlab = "WEIGHT (1000lb)", ylab
= "VS")
```

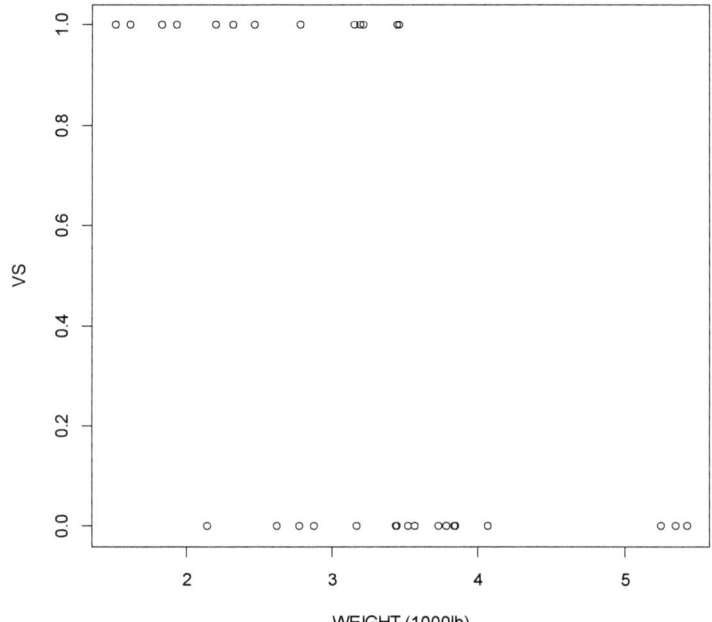

```
> cor(mtcars$vs, mtcars$wt) # pair-wise association
[1] -0.5549157

> model <- glm(formula= vs ~ wt , data=mtcars,
family=binomial)
> summary(model)
Call:
glm(formula = vs ~ wt, family = binomial, data = mtcars)
Deviance Residuals:
    Min      1Q   Median      3Q      Max
-1.9003  -0.7641  -0.1559   0.7223   1.5736

Coefficients:
            Estimate Std. Error z value Pr(>|z|)
(Intercept)   5.7147     2.3014    2.483  0.01302 *
wt           -1.9105     0.7279   -2.625  0.00867 **
---
Signif. codes:  0 '***' 0.001 '**' 0.01 '*' 0.05 '.' 0.1
' ' 1
(Dispersion parameter for binomial family taken to be 1)
    Null deviance: 43.860  on 31  degrees of freedom
Residual deviance: 31.367  on 30  degrees of freedom
AIC: 35.367
```

```
Number of Fisher Scoring iterations: 5
> ## CIs using profiled log-likelihood
> confint(model)
Waiting for profiling to be done...
                  2.5 %      97.5 %
(Intercept)   1.934013 11.2235317
wt           -3.636933 -0.7172072

> exp(coef(model)) ## odds ratios only
(Intercept)          wt
303.2876052   0.1479993
> exp(cbind(OR = coef(model), confint(model)))  ## odds
ratios and 95% CI
Waiting for profiling to be done...
                    OR        2.5 %        97.5 %
(Intercept) 303.2876052 6.91721352 7.487173e+04
wt            0.1479993 0.02633297 4.881136e-01

> newdata = data.frame(wt = 2.1)# imagine we have a 2100
lb car
> newdata
   wt
1 2.1
> predict(model, newdata, type="response")# what would be
the model's response for that car?
        1
0.8458651
####################################33
## test goodness of fit for the model
install.packages("ResourceSelection",
repos="http://cran.r-project.org")
library(ResourceSelection)
> hoslem.test(mtcars$vs, fitted(model))
        Hosmer and Lemeshow goodness of fit (GOF) test
data:  mtcars$vs, fitted(model)
X-squared = 7.8293, df = 8, p-value = 0.4503
```

```
######### plot ROC curve
install.packages("pROC",repos="http://cran.r-
project.org")
library(pROC)
> roc(vs ~ wt, data = mtcars)
Call:
roc.formula(formula = vs ~ wt, data = mtcars)
Data: wt in 18 controls (vs 0) > 14 cases (vs 1).
Area under the curve: 0.8413
> plot(roc(vs ~ wt, data = mtcars))
Call:
roc.formula(formula = vs ~ wt, data = mtcars)
Data: wt in 18 controls (vs 0) > 14 cases (vs 1).
Area under the curve: 0.8413
```

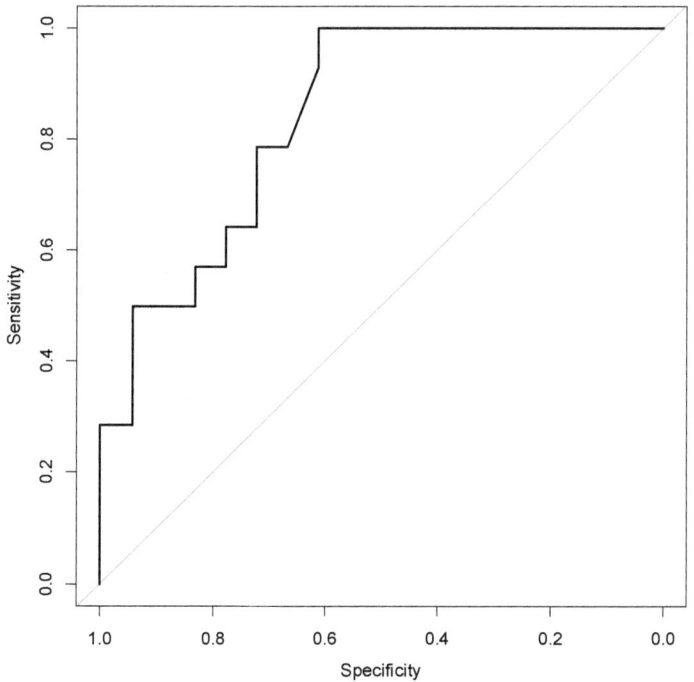

Interpretation: mtcars data lists 11 values for 32 cars. Because engine type is binary, applying the command summary(mtcars$vs) does not make sense. Instead, we use the tabulation command table(mtcars$vs) which shows that there are 18 v type and 14 s type cars. We wanted to build a logistic model that predicts the engine type (v or s) based on the car's weight. Note that vs and weight are negatively correlated (pair-wise association = -0.55).

Summary(model) output shows the coefficients, their standard errors, the z-statistic (sometimes called a Wald z-statistic), and their associated p-values. Weight is statistically significant (p-value < 0.05). The logistic regression coefficients give the change in the log odds (coefficients) of the outcome (vs) for a one unit (1000lb) increase in the predictor variable (wt). So for every 1000 pound change in car weight, the log odds of having a v engine (versus s engine) decreases by -1.9105. confint function to obtain confidence intervals for the coefficient estimates.

Let's interpret the odds ratio. Note that the odds ratio for the intercept is not generally interpreted. The odds ratio for weight is less than one which means that vs and weight are negatively correlated. In other words, the more a car weighs the less are its chances to have a v engine.

For a car that weighs 2,100lb, the model's predicted probability is 0.84.

Hosmer and Lemeshow goodness of fit test tests the hypothesis that the difference between the model and observed data (mtcars$vs) is statistically significant. Because the p-value of the test (0.4503) is greater than the significance level of 0.05, we reject the hypothesis, which means that our model fits the data well. More on this in the last chapter.

The receiver operating characteristic (ROC) curve plots the values of sensitivity (vertical axis) against one minus specificity, as the value of the cut-point is increased from 0 through to 1. It measures the model's discrimination ability or the ability to correctly classify s or v cars based on their weights. It has three possibilities/interpretations:

1) A diagonal line represents a model with no discrimination ability, or a worthless model. The more the model's ROC curve departs (to the left) from the diagonal line, the higher its discrimination ability (more area under the curve). This happens when the predictor is not a strong one.

2) A good model

3) An excellent model where the area between the diagonal line and the model is at its highest, which means that the predictor is a strong one. The following is a hypothetical ROC curve with labels of the three model possibilities. The bigger the area under the curve, the better the model:

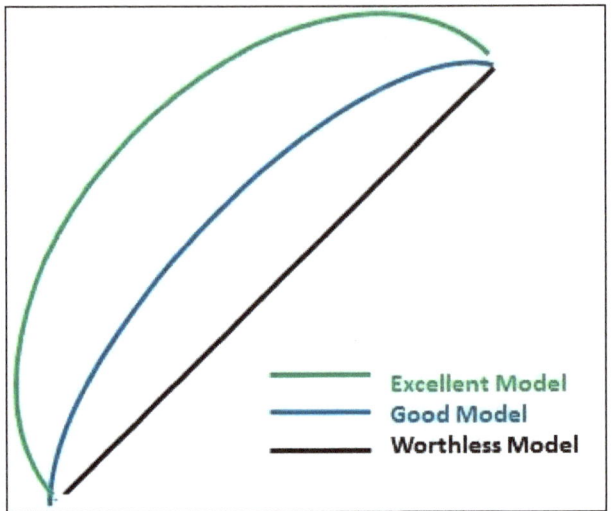

Example 2: Probit Regression

The probit regression is used when the outcome is binary, just like the logistic regression. It also gives similar inferences but the function (formula) is different (cumulative distribution function of the standard normal). This example will use the same data as in Example 1 but fit a probit model instead.

```
> model1 <- glm(formula= vs ~ wt , data=mtcars,
family=binomial(probit))
> summary(model1)

Call:
glm(formula = vs ~ wt, family = binomial(probit),
data = mtcars)

Deviance Residuals:
    Min         1Q    Median         3Q        Max
-1.90416   -0.75941  -0.08308    0.72889    1.57783

Coefficients:
            Estimate Std. Error z value Pr(>|z|)
(Intercept)   3.4793     1.2866    2.704  0.00684 **
wt           -1.1672     0.4086   -2.857  0.00428 **
---
Signif. codes:  0 '***' 0.001 '**' 0.01 '*' 0.05 '.' 0.1
' ' 1

(Dispersion parameter for binomial family taken to be 1)

    Null deviance: 43.860  on 31  degrees of freedom
Residual deviance: 31.141  on 30  degrees of freedom
AIC: 35.141

Number of Fisher Scoring iterations: 6
```

```
> confint(model1) # CI
Waiting for profiling to be done...
                 2.5 %      97.5 %
(Intercept)   1.230736   6.4119809
wt           -2.102847  -0.4565985
```

Interpretation: Weight is statistically significant (p-value < 0.05), and its coefficient of -1.1672 means that for every weight unit increase, the predicted probability of a v engine decreases by 1.1672.

Example 3: Relative Frequency Outcome

```
install.packages("MASS",repos="http://cran.r-
project.org")
library(MASS)
data(menarche)
> dim(menarche)
[1] 25   3
help(menarche)
```
Description
Proportions of female children at various ages during adolescence who have reached menarche.
Usage
menarche
Format
This data frame contains the following columns:
Age
Average age of the group. (The groups are reasonably age homogeneous.)
Total
Total number of children in the group.
Menarche
Number who have reached menarche.
```
> head(menarche)
    Age Total Menarche
1  9.21   376        0
2 10.21   200        0
3 10.58    93        0
4 10.83   120        2
5 11.08    90        2
6 11.33    88        5
> summary(menarche)
      Age            Total            Menarche
 Min.   : 9.21   Min.   :  88.0   Min.   :   0.00
 1st Qu.:11.58   1st Qu.:  98.0   1st Qu.:  10.00
 Median :13.08   Median : 105.0   Median :  51.00
 Mean   :13.10   Mean   : 156.7   Mean   :  92.32
 3rd Qu.:14.58   3rd Qu.: 117.0   3rd Qu.:  92.00
 Max.   :17.58   Max.   :1049.0   Max.   :1049.00


> cor(menarche) # correlation matrix
           Age      Total   Menarche
Age   1.0000000 0.3180055 0.6251757
Total 0.3180055 1.0000000 0.9185159
```

Menarche 0.6251757 0.9185159 1.0000000

```
plot(Menarche/Total ~ Age, data=menarche)
```

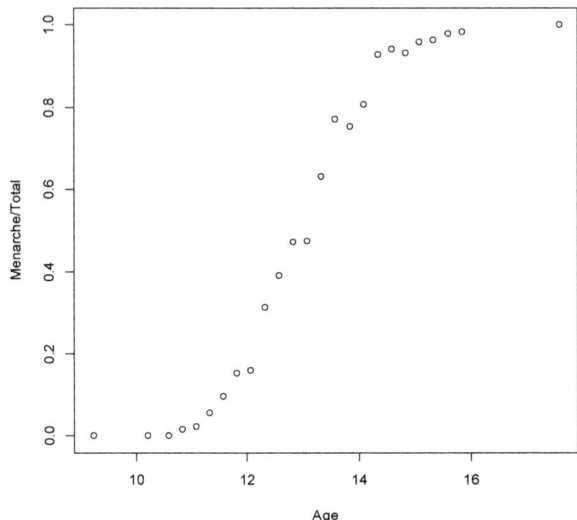

```
> menarche_total<- glm(cbind(Menarche, Total-Menarche) ~
Age, family=binomial, data=menarche)
> summary(menarche_total)

Call:
glm(formula = cbind(Menarche, Total - Menarche) ~ Age,
family = binomial,
    data = menarche)

Deviance Residuals:
    Min       1Q    Median       3Q       Max
-2.0363  -0.9953  -0.4900   0.7780    1.3675

Coefficients:
            Estimate Std. Error z value Pr(>|z|)
(Intercept) -21.22639    0.77068  -27.54   <2e-16 ***
Age           1.63197    0.05895   27.68   <2e-16 ***
---
Signif. codes:  0 '***' 0.001 '**' 0.01 '*' 0.05 '.' 0.1
' ' 1

(Dispersion parameter for binomial family taken to be 1)

    Null deviance: 3693.884  on 24  degrees of freedom
```

Residual deviance: 26.703   on 23   degrees of freedom
AIC: 114.76

Number of Fisher Scoring iterations: 4


```
plot(Menarche/Total ~ Age, data=menarche)
lines(menarche$Age, menarche_total$fitted, type="l",
col="red")
title(main="Rate of Menarche with Fitted Logistic
Regression Line")
```

############## Odds Ratio
```
> exp(cbind(OR = coef(menarche_total),
confint(menarche_total))) ## odds ratios and 95% CI
Waiting for profiling to be done...
                     OR          2.5 %          97.5 %
(Intercept) 6.046358e-10 1.270293e-10 2.610219e-09
Age         5.113931e+00 4.572696e+00 5.762244e+00

> anova(menarche_total, test="Chisq")
Analysis of Deviance Table
Model: binomial, link: logit
Response: cbind(Menarche, Total - Menarche)
Terms added sequentially (first to last)

      Df Deviance Resid. Df Resid. Dev  Pr(>Chi)
NULL                    24      3693.9
Age    1   3667.2        23        26.7 < 2.2e-16 ***
---
Signif. codes:  0 '***' 0.001 '**' 0.01 '*' 0.05 '.' 0.1
' ' 1
```

Rate of Menarche with Fitted Logistic Regression Line

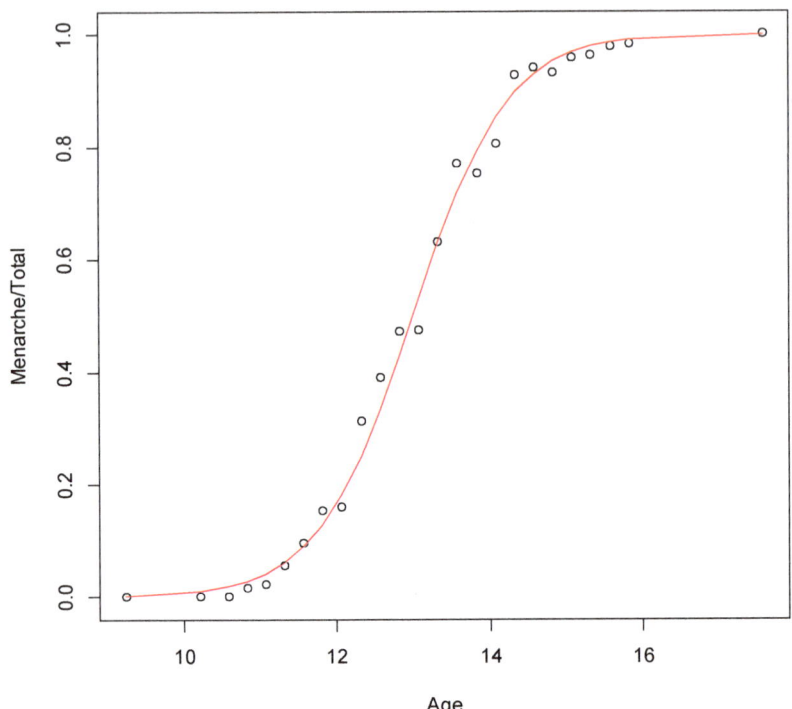

Interpretation: This example investigates three quantities about 25 female children groups: age of the group, total number of girls in the group, and the number of girls who have reached menarche within the group. Note the high positive correlation between Total and Menarche. Plotting the percent of those who reached menarche against age shows an exponential relationship. This suggests that a logistic fit could work. Age came out statistically significant (p-value < 0.05). And its OR (5.11) is greater than 1 which means that for every one year increase in age, the odds of having reached menarche increased by exp(1.632) = 5.11 times.

The chi square test of the model's goodness of fit gives a chi square value of 26.7 on 23 degrees of freedom with a p-value of zero (< 0.05). Hence, the null hypothesis that the model is significant is accepted.

Example 4: Multivariate Logistic Regression

This example uses the same dataset as in Example 1 (mtcars) but fits a model using more than one independent variable (predictor).

```
> data(mtcars)
```

```
> head(mtcars)
                   mpg cyl disp  hp drat    wt  qsec vs am gear carb
Mazda RX4         21.0   6  160 110 3.90 2.620 16.46  0  1    4    4
Mazda RX4 Wag     21.0   6  160 110 3.90 2.875 17.02  0  1    4    4
Datsun 710        22.8   4  108  93 3.85 2.320 18.61  1  1    4    1
Hornet 4 Drive    21.4   6  258 110 3.08 3.215 19.44  1  0    3    1
Hornet Sportabout 18.7   8  360 175 3.15 3.440 17.02  0  0    3    2
Valiant           18.1   6  225 105 2.76 3.460 20.22  1  0    3    1

> table(mtcars$vs)
 0  1
18 14
> summary(mtcars$wt) # car weight in 1000 lb)
   Min. 1st Qu.  Median    Mean 3rd Qu.    Max.
  1.513   2.581   3.325   3.217   3.610   5.424
> summary(mtcars$mpg) # Miles/(US) gallon
   Min. 1st Qu.  Median    Mean 3rd Qu.    Max.
  10.40   15.42   19.20   20.09   22.80   33.90

> cor(mtcars$vs, mtcars$wt) # pair-wise association
[1] -0.5549157
> cor(mtcars$vs, mtcars$mpg) # pair-wise association
[1] 0.6640389
> cor(mtcars$wt, mtcars$mpg) # pair-wise association
[1] -0.8676594

> round(cor(mtcars),2) # pair-wise association
       mpg   cyl  disp    hp  drat    wt  qsec    vs    am
gear  carb
mpg   1.00 -0.85 -0.85 -0.78  0.68 -0.87  0.42  0.66  0.60  0.48 -0.55
cyl  -0.85  1.00  0.90  0.83 -0.70  0.78 -0.59 -0.81 -0.52 -0.49  0.53
disp -0.85  0.90  1.00  0.79 -0.71  0.89 -0.43 -0.71 -0.59 -0.56  0.39
hp   -0.78  0.83  0.79  1.00 -0.45  0.66 -0.71 -0.72 -0.24 -0.13  0.75
drat  0.68 -0.70 -0.71 -0.45  1.00 -0.71  0.09  0.44  0.71  0.70 -0.09
wt   -0.87  0.78  0.89  0.66 -0.71  1.00 -0.17 -0.55 -0.69 -0.58  0.43
qsec  0.42 -0.59 -0.43 -0.71  0.09 -0.17  1.00  0.74 -0.23 -0.21 -0.66
vs    0.66 -0.81 -0.71 -0.72  0.44 -0.55  0.74  1.00  0.17  0.21 -0.57
am    0.60 -0.52 -0.59 -0.24  0.71 -0.69 -0.23  0.17  1.00  0.79  0.06
gear  0.48 -0.49 -0.56 -0.13  0.70 -0.58 -0.21  0.21  0.79  1.00  0.27
carb -0.55  0.53  0.39  0.75 -0.09  0.43 -0.66 -0.57  0.06  0.27  1.00
> model2 <- glm(formula= vs ~ wt + mpg , data=mtcars,
family=binomial)
> summary(model2)
Call:
glm(formula = vs ~ wt + mpg, family = binomial, data =
mtcars)

Deviance Residuals:
    Min       1Q   Median       3Q      Max
-2.2020  -0.5835  -0.2311   0.5376   1.7142
```

```
Coefficients:
            Estimate Std. Error z value Pr(>|z|)
(Intercept) -12.5412     8.4660  -1.481   0.1385
wt            0.5829     1.1845   0.492   0.6227
mpg           0.5241     0.2604   2.012   0.0442 *
---
Signif. codes:  0 `***' 0.001 `**' 0.01 `*' 0.05 `.' 0.1
` ' 1
(Dispersion parameter for binomial family taken to be 1)

    Null deviance: 43.860  on 31  degrees of freedom
Residual deviance: 25.298  on 29  degrees of freedom
AIC: 31.298
Number of Fisher Scoring iterations: 6

> exp(cbind(OR = coef(model2), confint(model2))) ## odds
ratios and 95% CI
Waiting for profiling to be done...
                    OR           2.5 %      97.5 %
(Intercept) 3.576157e-06 1.387865e-14 18.560698
wt          1.791153e+00 1.468780e-01 18.855651
mpg         1.688877e+00 1.097753e+00  3.207321>
> library(pROC)
Type 'citation("pROC")' for a citation.

Attaching package: `pROC'
The following objects are masked from `package:stats':
    cov, smooth, var

Warning message:
package `pROC' was built under R version 3.1.3
> roc(vs ~ wt + mpg, data = mtcars)
$wt

Call:
roc.formula(formula = vs ~ wt, data = mtcars)

Data: wt in 18 controls (vs 0) > 14 cases (vs 1).
Area under the curve: 0.8413

$mpg

Call:
roc.formula(formula = vs ~ mpg, data = mtcars)

Data: mpg in 18 controls (vs 0) < 14 cases (vs 1).
```

```
Area under the curve: 0.9107
```

Interpretation: Looking at the pair-wise associations, each predictor (wt and mpg) has high correlation with the outcome (vs), but there is high association between them (cor(mpg, wt) = -0.87), so one of them should have been excluded in order to avoid collinearity. We will ignore this issue here just for the sake of demonstrating how to fit a multivariate logistic regression. Looking at the estimates of model2, we now find that mpg is significant (p-value < 0.05) while weight is not. This is also translated in the OR of weight having a less than one 2.5% CI estimate, whereas mpg's OR CI is well above 1. This means that for every 1 US mile/gallon, the OR of having a v engine increases by exp(0.5241) = 1.68.

Comparing the area under the curve between the two models:

logit (vs ~ wt) Area under the curve: 0.8413

logit (vs ~ wt + mpg) Area under the curve: 0.9107

We can see that the second model is slightly better (larger area) than the first one.

Chapter 2: Multinomial Regression

This chapter discusses the case when the outcome/response has more than two values, and they might have an order or a weight. Example outcomes include SAT score, favorite food items, etc. Other names for this regression include polytomous, multiclass, multinomial logit, maximum entropy, or conditional maximum entropy.

Example 5: Multinomial Univariate Regression

This example investigates the four modes of fishing (beach, pier, boat and charter) in terms of catching rate. Library mnlogit was used to call the data Fish, library nnet was used to call the function multinom, and library AER was called to calculate p-values for the estimates because multinom does not provide them.

```
install.packages("mnlogit",           repos="http://cran.r-
project.org")
library(mnlogit)
> data(Fish)
> head(Fish)
            mode   income      alt   price catch chid
1.beach    FALSE 7083.332    beach 157.930 0.0678    1
1.boat     FALSE 7083.332     boat 157.930 0.2601    1
1.charter   TRUE 7083.332  charter 182.930 0.5391    1
1.pier     FALSE 7083.332     pier 157.930 0.0503    1
2.beach    FALSE 1250.000    beach  15.114 0.1049    2
2.boat     FALSE 1250.000     boat  10.534 0.1574    2
```

```
help(Fish)
Choice of Fishing Mode
Description
A data.frame containing data on choices of recreational
fishing mode. Data may depend on both the individual and
the alternative.
number of observations : 1182
country : United States
Usage
data(Fish)
Format
A dataframe containing :
mode - recreation mode choice, one of : beach, pier, boat
and charter
price - price for a mode for an individual
catch - fish catch rate for a mode for an individual
income - monthly income of an individual
chid - chooser ID: serial number of the individual

> table(Fish$alt)
  beach     boat charter     pier
   1182     1182    1182     1182
> dim(Fish)
[1] 4728      6

install.packages("nnet",                repos="http://cran.r-
project.org")
library(nnet) # in order to use multinom()
> test1 = multinom(mode ~ catch , data=Fish)
# weights:  3 (2 variable)
initial  value 3277.199870
final  value 2628.978646
converged
> summary(test1)
Call:
multinom(formula = mode ~ catch, data = Fish)

Coefficients:
              Values  Std. Err.
(Intercept) -1.2769131 0.04143854
catch        0.5514345 0.07032672

Residual Deviance: 5257.957
AIC: 5261.957
>  ##  extract  the  coefficients  from  the  model  and
exponentiate
```

```
> exp(cbind(OR = coef(test1), confint(test1))) ## odds
ratios and 95% CI
                  OR      2.5 %     97.5 %
(Intercept) 0.2788969 0.2571409 0.3024936
catch       1.7357412 1.5122471 1.9922654

## multinom from nnet does not calculate p-values, the
following does
install.packages("AER",                  repos="http://cran.r-
project.org")
library(AER)
coeftest(test1)
z test of coefficients:

             Estimate Std. Error z value  Pr(>|z|)
(Intercept) -1.276913   0.041439 -30.815 < 2.2e-16 ***
catch        0.551435   0.070327   7.841 4.468e-15 ***
---
Signif. codes:  0 '***' 0.001 '**' 0.01 '*' 0.05 '.' 0.1
' ' 1

### what about the multinomial aspect or modes of
fishing?
> Fish$grp2 = relevel(as.factor(Fish$alt), ref ="beach")
> test2<- multinom(grp2 ~ catch, data=Fish)
# weights:  12 (6 variable)
initial  value 6554.399739
iter  10 value 6067.792670
final  value 6067.416936
converged
> summary(test2)
Call:
multinom(formula = grp2 ~ catch, data = Fish)

Coefficients:
        (Intercept)      catch
boat      0.2717424 -1.339993
charter  -0.7768129  2.078085
pier      0.3116005 -1.579104

Std. Errors:
        (Intercept)      catch
boat     0.05536667 0.1859031
charter  0.06262493 0.1405729
pier     0.05532657 0.1917634

Residual Deviance: 12134.83
```

```
AIC: 12146.83
> coeftest(test2)
z test of coefficients:

                      Estimate Std. Error   z value   Pr(>|z|)
boat:(Intercept)      0.271742   0.055367    4.9080 9.199e-07 ***
boat:catch           -1.339993   0.185903   -7.2080 5.677e-13 ***
charter:(Intercept)  -0.776813   0.062625  -12.4042 < 2.2e-16 ***
charter:catch         2.078085   0.140573   14.7830 < 2.2e-16 ***
pier:(Intercept)      0.311600   0.055327    5.6320 1.781e-08 ***
pier:catch           -1.579104   0.191763   -8.2346 < 2.2e-16 ***
---
Signif. codes:  0 '***' 0.001 '**' 0.01 '*' 0.05 '.' 0.1 ' ' 1
```

Interpretation: The outcome/response/dependent variable takes more than two options (hence multinomial): beach, pier, boat, and charter. These options are not weighted or ordered. The model test attempts to explain the four choices based on catching rate. Since there is only one predictor, this is a univariate regression. The p-value for catch is less than 0.05 therefore it is significant. This is also expressed in the OR having a more than one value of 1.73 and both CI values are greater than 1. This OR means that for every one unit increase in catching rate the log odds of fishing mode is increased by 0.55.

However, this output did not differentiate between the four groups or modes of fishing. In order to do so we first need to explicitly decide which group/category is our baseline or reference. Let's assume we want to compare how the catching rate (catch) affects the mode of fishing considering beach as our reference group. In model test2, each estimate is in contrast to the reference group of beach. Every unit increase in catch decreases the log odds of boating mode by 1.3, increases the log odds of charter mode by 2.07, and decreases the log odds of pier mode by 1.57. More on model test2 in the next example.

Example 6: Multinomial Multivariate Regression

This example uses the same data and libraries as in Example 4 but attempts to explain the mode of fishing based on more than one predictor (multivariate): catching rate and price of fishing mode. It uses the same libraries: mnlogit, nnet, AER.

```
> test3 <- multinom(grp2 ~ catch + price, data=Fish)
# weights:  16 (9 variable)
initial  value 6554.399739
iter  10 value 5992.481341
final  value 5916.419012
converged
```

```
> summary(test3)
Call:
multinom(formula = grp2 ~ catch + price, data = Fish)
Coefficients:
          (Intercept)        catch          price
boat        0.8424511  -0.9351842   -0.008842989
charter    -0.5207365   2.2344963   -0.003210052
pier        0.2556874  -1.6998144    0.000774552

Std. Errors:
          (Intercept)        catch          price
boat       0.06996072  0.1909751    0.0006758266
charter    0.07408603  0.1487172    0.0005009787
pier       0.06610705  0.1993928    0.0004247535
Residual Deviance: 11832.84
AIC: 11850.84

> coeftest(test3)
z test of coefficients:
                        Estimate   Std. Error   z value   Pr(>|z|)
boat:(Intercept)       0.84245112   0.06996072   12.0418 < 2.2e-16  ***
boat:catch            -0.93518417   0.19097512   -4.8969 9.737e-07  ***
boat:price            -0.00884299   0.00067583  -13.0847 < 2.2e-16  ***
charter:(Intercept)   -0.52073647   0.07408603   -7.0288 2.083e-12  ***
charter:catch          2.23449633   0.14871723   15.0251 < 2.2e-16  ***
charter:price         -0.00321005   0.00050098   -6.4076 1.479e-10  ***
pier:(Intercept)       0.25568743   0.06610705    3.8678 0.0001098  ***
pier:catch            -1.69981444   0.19939282   -8.5250 < 2.2e-16  ***
pier:price             0.00077455   0.00042475    1.8235 0.0682227  .
---
Signif. codes:  0 '***' 0.001 '**' 0.01 '*' 0.05 '.' 0.1 ' ' 1
```

Interpretation: The outcome/response/dependent variable takes more than two options (hence multinomial): beach, pier, boat, and charter. These options are not weighted or ordered. The model test attempts to explain the four choices based on catching rate and price. Since there is more than one predictor, this is a multivariate regression. P-values for all predictors except for price's effect on pier, are zero (< 0.05) hence these are the significant ones. Moreover, test2 AIC (12146.83) is greater than test's AIC of test 3 11850.84. Therefore test3 is better than test2. The table below interprets the coefficients of test3:

Independent – dependent group coefficient	Interpretation	Significant? (p-value < 0.05?)
Catch – boat -0.9351842	Every unit increase in catching rate decreases the log odds of boating by 0.9	Yes
Catch – charter 2.2344963	Every unit increase in catching rate increases the log odds of boating by 2.2	Yes
Catch – pier -1.6998144	Every unit increase in catching rate decreases the log odds of boating by 1.7	Yes
Price – boat -0.008842989	Every unit increase in price decreases the log odds of boating by 0.009	Yes
Price – charter -0.003210052	Every unit increase in price decreases the log odds of chartering by 0.003	Yes
Price – pier 0.000774552	Every unit increase in price increases the log odds of fishing on a pier by 0.0008	No

Chapter 3: Ordinal Regression

Example 7: Ordinal Logistic Regression

This example uses the housing data from MASS library (Frequency Table from a Copenhagen Housing Conditions Survey). In this example we fit a model (house.plr) which estimates residents satisfaction based on type of accommodation, but each case will be weighted/scored/ordered using the number of residents in each class.

```
> library(MASS) # for polr
data(housing)
help(housing)
Description
The housing data frame has 72 rows and 5 variables.
Usage
housing
Format
Sat: Satisfaction of householders with their present
housing circumstances, (High, Medium or Low, ordered
factor).

Infl: Perceived degree of influence householders have on
the management of the property (High, Medium, Low).

Type: Type of rental accommodation, (Tower, Atrium,
Apartment, Terrace).

Cont: Contact residents are afforded with other
residents, (Low, High).
```

Freq: Frequencies: the numbers of residents in each class.

```
> head(housing)
     Sat    Infl  Type Cont Freq
1    Low    Low Tower  Low   21
2 Medium    Low Tower  Low   21
3   High    Low Tower  Low   28
4    Low Medium Tower  Low   34
5 Medium Medium Tower  Low   22
6   High Medium Tower  Low   36

> house.plr <- polr(Sat ~ Type , weights = Freq, data =
housing)
> house.plr
Call:
polr(formula = Sat ~ Type, data = housing, weights =
Freq)

Coefficients:
TypeApartment     TypeAtrium     TypeTerrace
   -0.4684229     -0.2803333      -1.0534395

Intercepts:
  Low|Medium  Medium|High
-1.118228135  0.003082102

Residual Deviance: 3594.803
AIC: 3604.803

> summary(house.plr, digits = 3)

Re-fitting to get Hessian

Call:
polr(formula = Sat ~ Type, data = housing, weights =
Freq)

Coefficients:
              Value Std. Error t value
TypeApartment -0.468      0.116   -4.04
TypeAtrium    -0.280      0.150   -1.86
TypeTerrace   -1.053      0.147   -7.14

Intercepts:
           Value  Std. Error t value
Low|Medium -1.118      0.100  -11.223
```

```
Medium|High    0.003    0.096        0.032

Residual Deviance: 3594.803
AIC: 3604.803

library(AER)
> coeftest(house.plr)
Re-fitting to get Hessian
z test of coefficients:
              Estimate Std. Error   z value  Pr(>|z|)
TypeApartment -0.4684229  0.1160362  -4.0369 5.417e-05 ***
TypeAtrium    -0.2803333  0.1503585  -1.8644   0.06226 .
TypeTerrace   -1.0534395  0.1474880  -7.1425 9.162e-13 ***
Low|Medium    -1.1182281  0.0996390 -11.2228 < 2.2e-16 ***
Medium|High    0.0030821  0.0956504   0.0322   0.97429
---
Signif. codes:  0 '***' 0.001 '**' 0.01 '*' 0.05 '.' 0.1 ' ' 1
```

Interpretation: The table below interprets the 3 coefficients of

house.plr:

Independent – dependent group coefficient	Interpretation	Significant? (p-value < 0.05?)
Satisfaction – apartment -0.468	Compared to living in a tower, living in an apartment decreases the log odds of satisfaction by 0.468.	Yes
Satisfaction – atrium -0.280	Compared to living in a tower, living in house with an atrium decreases the log odds of satisfaction by 0.280.	No
Satisfaction – Terrace -1.053	Compared to living in a tower, living in house with a terrace decreases the log odds of satisfaction by 1.053.	Yes

 The End